Dimensions Math
Tests 4A

Singapore Math Inc.

Published by Singapore Math Inc.

19535 SW 129th Avenue
Tualatin, OR 97062
www.singaporemath.com

Dimensions Math® Tests 4A
ISBN 978-1-947226-53-1

First published 2019
Reprinted 2020

Copyright © 2019 by Singapore Math Inc.
All rights reserved. This book or any portion thereof may not be reproduced or used in any manner whatsoever without the express written permission of the publisher.

Printed in China

Acknowledgments

Editing by the Singapore Math Inc. team.
Design and illustration by Cameron Wray with Carli Fronius.

Preface

Dimensions Math® Tests is a series of assessments to help teachers systematically evaluate student progress. The tests align with the content of Dimensions Math K–5 textbooks.

Dimensions Math Tests K uses pictorially engaging questions to test student ability to grasp key concepts through various methods including circling, matching, coloring, drawing, and writing numbers.

Dimensions Math Tests 1–5 have differentiated assessments. Tests consist of multiple-choice questions that assess comprehension of key concepts, and free response questions for students to demonstrate their problem-solving skills.

Test A focuses on key concepts and fundamental problem-solving skills.

Test B focuses on the application of analytical skills, thinking skills, and heuristics.

Contents

Chapter	Test	Page
Chapter 1 **Numbers to One Million**	Test A	1
	Test B	7
Chapter 2 **Addition and Subtraction**	Test A	13
	Test B	19
Chapter 3 **Multiples and Factors**	Test A	25
	Test B	31
Chapter 4 **Multiplication**	Test A	37
	Test B	43
Chapter 5 **Division**	Test A	49
	Test B	57
	Continual Assessment 1 Test A	65
	Continual Assessment 1 Test B	75
Chapter 6 **Fractions**	Test A	85
	Test B	91

Chapter	Test	Page
Chapter 7 **Adding and Subtracting Fractions**	Test A Test B	97 103
Chapter 8 **Multiplying a Fraction and a Whole Number**	Test A Test B	109 115
Chapter 9 **Line Graphs and Line Plots**	Test A Test B **Continual Assessment 2 Test A** **Continual Assessment 2 Test B**	121 129 137 147
Answer Key		157

BLANK

Name: _____

Date: _____

20 min   Score

30

Test A

Chapter 1 Numbers to One Million

Section A (2 points each)
Circle the correct option: **A**, **B**, **C**, or **D**.

1 1 million is _____ thousands.

    A 1,000   **B** 10

    C 10,000   **D** 100

2 In 572,342, the digit 7 stands for 7 _____.

    A hundred thousands   **B** ten hundreds

    C ten thousands   **D** thousands

Chapter 1 Test A   1

3 In 720,150, the value of the digit 7 is _____.

A 7

B 700,000

C 700

D 70,000

4 150,000 + 270,000 = ☐

A 120,000

B 42,000

C 420,000

D 320,000

5 98,415 is greater than _____.

A 98,514

B 98,541

C 98,451

D 98,145

Section B (2 points each)

Use the information in the place-value chart to answer questions 6–9.

Ten Thousands	Thousands	Hundreds	Tens	Ones
5	9	0	1	6

6 What number is 10,000 more than this number? ☐

7 What number is 1,000 less than this number? ☐

8 Write the number in words.

_____

9 Write this number in expanded form.

_____

Chapter 1 Test A

Use the number line to answer questions 10–12.

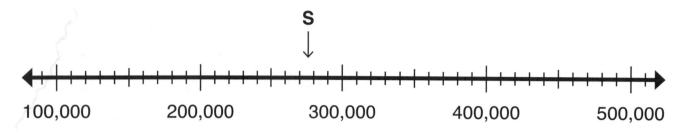

10 Circle the number that best corresponds to the number indicated by S.

| 120,020   277,352   303,198   252,500 |

11 The number indicated by S is _____ when rounded to the nearest hundred thousand.

12 _____ is halfway between 400,000 and 500,000.

13 Complete the number pattern.

| 136,263 | 137,264 | | | 140,267 | |

14 Write the numbers in order from least to greatest.

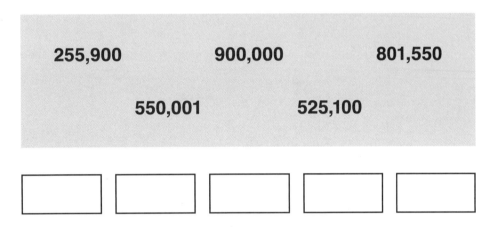

15 A grand piano costs 5 times as much as an upright piano. If the grand piano costs $20,000, how much does the upright piano cost?

BLANK

Name: _____

Date: _____

20 min Score

30

Test B

Chapter 1 Numbers to One Million

Section A (2 points each)
Circle the correct option: **A**, **B**, **C**, or **D**.

1 In 875,742, the digit 8 stands for 8 _____.

   A thousands

   B hundred tens

   C hundred thousands

   D hundred hundreds

2 In 492,675, the value of the digit 4 is equal to 40 _____.

   A hundred thousands

   B hundreds

   C thousands

   D ten thousands

3 1,000,000 is _____ ten thousands.

    A 10   **B** 1,000

    C 100   **D** 10,000

4 _____ is 10,000 more than the quotient of 630,000 ÷ 9.

    A 10 ten thousands   **B** 80 tens

    C 8 ten thousands   **D** 8 hundred thousands

5 What number when rounded to the nearest thousand and the nearest ten thousand is 310,000?

    A 309,500   **B** 299,000

    C 310,900   **D** 309,199

Section B (2 points each)

Use the information in the place-value chart to answer questions 6–9.

Hundred Thousands	Ten Thousands	Thousands	Hundreds	Tens	Ones
3	1	0	6	8	2

6 What number is 100,000 more than this number? ☐

7 What number is 10,000 less than this number? ☐

8 In this number, the digit 0 stands for 0 _____, and its value is _____.

9 Write this number in expanded form.

_____

Use the number line to answer questions 10–12.

10 Circle the number that best corresponds to the number indicated by N.

| 980,137   903,100   930,925   900,030 |

11 The number indicated by N is _____ when rounded to the nearest hundred thousand.

12 _____ is halfway between 900,000 and 1,000,000.

13 What is the least 6-digit odd number that can be formed with these digits? Use each digit only once.

3, 1, 8, 0, 5, 9

_____

14 There are 357,249 students in School District A, 352,058 students in School District B, 326,211 students in School District C, and 633,621 students in School District D. Arrange the school districts in order from the least to the greatest number of students.

_____

15 The price of a scooter is $2,000 and the price of a motorcycle is $3,000. A shop sold 6 scooters and a motorcycle. How much did the shop receive in total for the scooters and motorcycle?

BLANK

Name: _____

Date: _____

25 min   Score

30

Test A

Chapter 2 Addition and Subtraction

Section A (2 points each)
Circle the correct option: **A**, **B**, **C**, or **D**.

1 What is the estimate for 71,078 + 18,805 when each number is rounded to the nearest ten thousand?

  A 75,000   B 90,000

  C 80,000   D 50,000

2 10,000 = 2,810 + ☐

  A 7,290   B 12,810

  C 8,290   D 7,190

3 20,000 − 672 = ☐

    A 19,328   **B** 13,280

    C 19,438   **D** 20,672

4 40,891 + 8,510 = ☐

    A 32,381   **B** 48,301

    C 49,401   **D** 48,401

5 3,821 − 2,999 = ☐

    A 821   **B** 822

    C 1,821   **D** 6,820

Section B (2 points each)

6 Circle the number that is equal to 38,753 + 51,335. Use estimation.

| 110,988 | 10,352 | 90,088 | 80,808 |

7 Circle the number that is equal to 741,054 − 46,281 without calculating the exact number.

| 300,773 | 694,773 | 780,085 | 700,009 |

8 65,300 − 49 = ☐

9 238,070 + 43,878 = ☐

10 85,208 − ☐ = 78,172

11 A farmer sold 4,216 chicken eggs and 695 duck eggs last year. How many eggs did he sell altogether?

Section C (4 points each)

12 Laila bought a scooter for $1,250. She also bought a helmet that cost $1,125 less than the scooter. How much did she pay for the scooter and the helmet altogether?

13 There were 12,000 people at a baseball game on Friday. 5,500 of them were men. There were 100 more women than men at the game and the rest were children. How many children were at the game?

Name: _____

Date: _____

25 min   Score

30

Test B

Chapter 2 Addition and Subtraction

Section A (2 points each)
Circle the correct option: **A**, **B**, **C**, or **D**.

1 What is the estimate for 875 + 8,750 + 87,500 when each number is rounded to the nearest thousand?

    A 120,000   **B** 98,000

    C 90,000   **D** 88,000

2 8,470 + ☐ = 10,000

    A 1,530   **B** 2,630

    C 1,630   **D** 18,470

3 77,000 − 7,009 = ☐

　A 77,009　　　　　　　　B 70,991

　C 69,991　　　　　　　　D 69,001

4 31,670 − ☐ = 998

　A 30,668　　　　　　　　B 30,672

　C 31,672　　　　　　　　D 30,562

5 899 + 798 + 697 + 596 = ☐

　A 2,990　　　　　　　　B 299

　C 3,000　　　　　　　　D 2,090

Section B (2 points each)

6 Circle the number that is equal to 6,831 + 71,582 + 1,984 without calculating the exact number.

| 75,408 | 80,397 | 100,397 | 79,400 |

7 Circle the number that is equal to 320,176 − 28,524 without calculating the exact number.

| 200,372 | 338,562 | 290,600 | 291,652 |

8 59,735 − 28 = ☐

9 438,650 − ☐ = 71,576

10 75,008 + 127,124 = ☐

11 Cody had $3,582. He bought a guitar and has $3,356 left. How much did the guitar cost?

Section C (4 points each)

12 8,152 people were at the State Fair on Friday. 1,250 more people were at the State Fair on Saturday than on Friday. 899 fewer people were at the State Fair on Sunday than on Saturday. How many people attended the State Fair altogether on Saturday and Sunday?

13 A department store had 2,856 shirts in the warehouse. There were 2,002 fewer shirts at the store than in the warehouse. Some shirts were transferred from the warehouse to the store and now the store has 1,449 shirts. How many shirts were transferred from the warehouse to the store?

Name: _____

Date: _____

25 min   Score

30

Test A

Chapter 3 Multiples and Factors

Section A (2 points each)
Circle the correct option: **A**, **B**, **C**, or **D**.

1 What is the 12th multiple of 5?

  A 50 **B** 60

  C 120 **D** 70

2 48 is a common multiple of _____.

  A 3 and 9 **B** 4 and 7

  C 5 and 12 **D** 3 and 12

Chapter 3 Test A   25

3 _____ is not a factor of 50.

A 2

B 10

C 4

D 5

4 Which one of these numbers is a prime number?

A 22

B 39

C 41

D 34

5 _____ is a common factor of 12 and 18.

A 6

B 4

C 12

D 9

Section B (2 points each)

6 Circle the numbers that are multiples of 8.

| 7 16 22 28 32 40 44 |

7 List the first five multiples of 6.

_____

8 List the first two common multiples of 2, 3, and 5.

_____

9 Circle the numbers that are factors of 36.

| 2 5 6 8 9 10 12 |

10 Circle the numbers that are prime numbers.

| 30 | 17 | 5 | 27 | 14 | 11 |

11 Circle the numbers that have 4 as a factor.

| 2 | 8 | 10 | 16 | 40 |

12 What is the greatest common factor of 18 and 36?

_____

13 Find the sum of the first four multiples of 7.

_____

14 The Nature Train at a park leaves the station every 5 minutes. The Jungle Train leaves the same station every 8 minutes. The two trains leave the station at 9:00 a.m. What time will the two trains leave the station at the same time again?

15 A baker has 30 chocolate chip cookies and 24 lemon cookies. He wants to put all the cookies in boxes so that each box will have the same number of each cookie. What is the greatest number of boxes he can make?

BLANK

Name: _____

Date: _____

25 min   Score

30

Test B

Chapter 3 Multiples and Factors

Section A (2 points each)
Circle the correct option: **A**, **B**, **C**, or **D**.

1 What is the 15th multiple of 9?

    A 108   **B** 45

    C 135   **D** 159

2 Which one is a common multiple of 5 and 6?

    A 36   **B** 60

    C 15   **D** 40

3 Which number has 8 as a factor?

    A 32   **B** 81

    C 42   **D** 54

4 _____ is neither a prime nor a composite number.

    A 7   **B** 10

    C 11   **D** 1

5 Which number has 3 and 8 as common factors?

    A 60   **B** 36

    C 72   **D** 16

Section B (2 points each)

6 List the first ten multiples of 9.

_____

7 Find the first two common multiples of 4, 6, and 8.

_____

8 List the factors of 66 in order from least to greatest.

_____

9 Check (✓) the statements that are true.

A prime number is always an odd number.	
A composite number has more than 2 factors.	
A multiple of an even number is always even.	

10 List the prime numbers that are between 10 and 20.

_____

11 Find the missing factor.

72 = 3 × 3 × ☐

12 Circle the number that has exactly 4 factors.

| 22　5　2　9 |

13 Circle the number that has a factor of 3 and is also a multiple of 6.

| 9　21　16　36　27 |

14 Three buses leave a bus station at intervals of 4, 6, and 8 minutes. The buses all leave at 7:00 a.m. When do the three buses next leave the station at the same time?

15 Sara has 60 stickers, 72 chocolate bars, and 96 bubble bottles. She wants to put together party bags that have the same combination of these items. What is the greatest number of bags she can make?

BLANK

Name: _____

Date: _____

30 min   Score

40

Test A

Chapter 4 Multiplication

Section A (2 points each)
Circle the correct option: **A**, **B**, **C**, or **D**.

1 15,000 × 3 = ☐

    A 4,500   **B** 45,000

    C 450,000   **D** 30,000

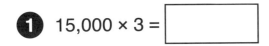

2 6 × 199 = ☐

    A 1,200   **B** 1,206

    C 1,194   **D** 594

Chapter 4 Test A   37

3 700 × 20 = ☐

A 14,000

B 70,000

C 6,400

D 140,000

4 12,624 × 3 = ☐

A 7,872

B 36,862

C 37,872

D 36,872

5 857 × 88 = ☐

A 75,316

B 74,416

C 68,416

D 75,416

Section B (2 points each)

6 Use mental calculation to find the product.

5 × 810 = ☐

Use the following expression to answer questions 7–9.

2,214 × 3

7 Sofia estimated the product to be 6,600. With what number did she replace 2,214?

8 Alex estimated the product to be 6,900. With what number did he replace 2,214?

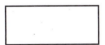

9 Whose estimate is closer to the actual product?

1,000

10 Circle the number that is equal to 32,674 × 4 without calculating the exact product.

| 13,096 | 130,696 | 1,369 | 130,575 |

11 531 × 6 = ☐

12 16 × 12 = ☐

13 Write > or < in the ◯. Use estimation.

39 × 73 ◯ 733 × 40

14 One airplane ticket costs $313.
How much do 6 tickets cost?

Section C (4 points each)

15 A pair of skis cost $281. A snow mobile cost 7 times as much as the pair of skis. Tom bought 2 pairs of skis and a snow mobile. How much did he pay in total for the skis and the snow mobile?

16 920 yellow roses were used in a rose festival parade. Twice as many white roses as yellow roses were used. Three times as many red roses as white roses were used. How many white and red roses were used altogether?

A bakery makes 1,296 chocolate croissants and 3,888 plain croissants daily. How many chocolate and plain croissants does the bakery make in 15 days?

Name: _____

Date: _____

30 min   Score

40

Test B

Chapter 4 Multiplication

Section A (2 points each)
Circle the correct option: **A**, **B**, **C**, or **D**.

1 1,999 × 9 = ☐

    A 179,991   **B** 18,000

    C 17,991   **D** 18,009

2 50 × 463 = ☐

    A 2,315   **B** 23,050

    C 20,150   **D** 23,150

3 6,014 × 7 = ☐

A 42,098

B 42,078

C 42,908

D 42,008

4 878 × 32 = ☐

A 28,016

B 26,340

C 28,096

D 26,100

5 How many minutes are there in 30 hours? *There are 60 minutes in one hour.*

A 300

B 1,800

C 180

D 3,600

Section B (2 points each)

6 Use mental calculation to find the product.

5,300 × 8 = ☐

Use the following expression to answer questions 7—9.

64,489 × 5

7 Mei estimated the product to be 325,000. With what number did she replace 64,489?

☐

8 Dion estimated the product to be 320,000. With what number did he replace 64,489?

☐

9 Whose estimate is closer to the actual product? Why?

10 Circle the number that is equal to 319 × 81 without calculating the exact product.

| 24,009 | 25,839 | 2,839 | 25,820 |

11 5 × 17,742 = ☐

12 22 × 93 = ☐

13 Write > or < in the ◯. Use estimation.

5 × 7,137 ◯ 959 × 15

14 Estimate to arrange the following expressions in order from least to greatest.

| 32,185 × 2 | 9,039 × 4 | 10,999 × 3 | 5,500 × 9 |

Section C (4 points each)

15 Amy saved $350 in January. She saved twice as much in February. She saved 6 times as much the rest of the year as she saved in February. How much more did she save the rest of the year than in January and February combined?

16 A machine in a factory wraps 116 chocolate bars in one minute. How many chocolate bars will 6 machines wrap in 10 minutes?

A day pass at an amusement park costs $55 per person. A night pass at the same park costs $36. The park sold 918 day passes and 354 night passes yesterday. How much money did the park receive from selling the day and night passes altogether?

Name: _____

Date: _____

30 min   Score

40

Test A

Chapter 5 Division

Section A (2 points each)
Circle the correct option: **A**, **B**, **C**, or **D**.

1 8,100 ÷ 9 = ☐

    A 900   **B** 800

    C 90   **D** 9,000

2 Which of the following gives the greatest quotient? Use estimation.

    A 963 ÷ 9   **B** 910 ÷ 8

    C 894 ÷ 5   **D** 478 ÷ 7

Chapter 5 Test A   49

3 Which of the following is equal to 3,504 ÷ 6? Use estimation.

   A 876 **B** 584

   C 684 **D** 494

4 Which of the following gives a quotient of 62?

   A 234 ÷ 9 **B** 1,240 ÷ 2

   C 554 ÷ 5 **D** 436 ÷ 7

5 Which of the following gives a remainder of 3?

   A 1,002 ÷ 3 **B** 9,203 ÷ 4

   C 224 ÷ 6 **D** 993 ÷ 8

Section B (2 points each)

6 Use mental calculation to find the value of 950 ÷ 5.

☐

7 708 ÷ 5 is _____ with a remainder of _____.

8 Mei estimated the quotient for an expression to be 200. Which expression is she estimating? Circle it.

| 1,528 ÷ 3   848 ÷ 4   2,015 ÷ 7 |

9 Write > or < in the ◯. Use estimation.

6,493 ÷ 8 ◯ 2,915 ÷ 3

10 4,941 ÷ 2 is _____ with a remainder of _____.

11 5,332 ÷ 9 is _____ with a remainder of _____.

12 Circle the number that 1,507 ÷ 7 is closest to.

| 300 | 200 | 250 | 750 |

13 What is the number of boxes needed for 312 cookies if each box holds 4 cookies?

14 Five friends are going to share the cost of renting a cabin equally. The rental is $1,595. How much will each friend pay?

Section C (4 points each)

15 A bakery delivers a total of 2,176 cookies equally to 8 coffee shops daily. How many cookies does the bakery deliver to each coffee shop in one week?

16 A teacher has 600 counters. He keeps 96 counters at his desk and gives the rest to 3 groups of students. The first group of students gets twice as many counters as the second group of students. The third group of students gets four times as many counters as the second group. How many counters did the second group of students get?

17 Three friends pay $1,025 to rent a vacation home. Beth contributed $80 more than Danny. Katie contributed $20 less than Beth. How much did Danny contribute?

Name: _____

Date: _____

30 min Score

40

Test B

Chapter 5 Division

Section A (2 points each)
Circle the correct option: **A**, **B**, **C**, or **D**.

1 49,000 ÷ 7 = ☐

  A 700   **B** 70,000

  C 70   **D** 7,000

2 Which of the following gives a quotient greater than 900 ÷ 5? Use estimation.

  A 1,011 ÷ 7   **B** 449 ÷ 3

  C 8,840 ÷ 40   **D** 1,600 ÷ 10

3 705 ÷ 3 is _____ with a remainder of 0.

    A 135   **B** 205

    C 235   **D** 201

4 Which of the following gives the greatest quotient?

    A 996 ÷ 5   **B** 847 ÷ 3

    C 1,205 ÷ 6   **D** 1,600 ÷ 8

5 Which of the following gives the greatest remainder?

    A 657 ÷ 7   **B** 1,385 ÷ 2

    C 5,000 ÷ 4   **D** 464 ÷ 9

Section B (2 points each)

6 Use mental calculation to find the value of 15,000 ÷ 6.

Use the following expression to answer questions 7–9.

5,901 ÷ 7

7 Dion estimated the quotient to be 800. With what number did he replace 5,901?

8 Alex estimated the quotient to be 900. With what number did he replace 5,901?

9 Whose estimate will be closer to the actual product? Why?

10 333 ÷ 8 is _____ with a remainder of _____.

11 8,229 ÷ 5 is _____ with a remainder of _____.

12 Write > or < in the ◯.

18,000 ÷ 2 ◯ 2 × 4,538

13 A shop sold 2 scooters for $3,090. One of the scooters cost twice as much as the other scooter. How much was the cheaper scooter?

14 Some teachers and 115 fourth graders are going on a field trip in 5 buses. Each bus will have the same number of students and 2 teachers. How many students and teachers will be on each bus altogether?

Section C (4 points each)

15 A baseball coach bought a bucket of baseballs, 4 baseball bats, and 3 baseball gloves for $274. The bucket of baseballs cost $120 and each bat cost $7 more than each glove. How much did the coach pay for each bat?

16 Three machines in a factory can wrap 8,181 pieces of candy in 9 minutes. How many pieces of candy can one machine wrap in 1 minute?

17 A cruise ship had 3 times as many adult passengers as child passengers when it left its first port. At a new port, 116 adults and 50 children boarded the ship. There are now 1,216 more adults than children on the ship. How many children were on the ship at first?

Name: _____

Date: _____

50 min   Score

60

Test A

Continual Assessment 1

Section A (2 points each)
Circle the correct option: **A**, **B**, **C**, or **D**.

1 In 32,048 the digit _____ is in the thousands place.

  A 3   B 2

  C 0   D 4

2 7,000 × 6 = ☐

  A 42,000   B 4,200

  C 52,000   D 420,000

3 Find the sum of 374,224 and 89,062.

  A 463,286   B 503,286

  C 453,226   D 460,386

4 30,000 − 684 = ☐

　　A 29,684　　　　　　　　B 24,160

　　C 28,330　　　　　　　　D 29,316

5 Which one is a common multiple of 2, 6, and 9?

　　A 16　　　　　　　　　　B 18

　　C 12　　　　　　　　　　D 27

6 Which one is a factor of 48?

　　A 15　　　　　　　　　　B 12

　　C 18　　　　　　　　　　D 7

7 30,000 × 6 = ☐

　　A 180,000　　　　　　　B 18,000

　　C 90,000　　　　　　　　D 160,000

8 3,075 × 9 = ☐

A 27,075

B 28,032

C 27,675

D 26,775

9 Use mental calculation to find the quotient of 35,000 ÷ 7.

A 4,000

B 500

C 6,000

D 5,000

10 9,876 ÷ 4 is closest to which number? Use estimation.

A 250

B 2,000

C 3,000

D 2,500

Section B (2 points each)

11 Count on by ten thousand to complete the number pattern.

| 91,489 | | | | |

12 Mark the approximate location of 651,264 on the number line.
Then round 651,264 to the nearest hundred thousand.

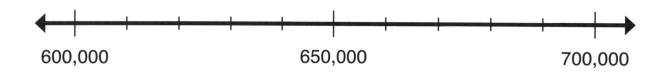

651,264 is _____ when rounded to the nearest hundred thousand.

13 ⬚ = 600 + 5 + 50,000 + 3,000 + 20

14 Write >, <, or = in the ◯.

600 thousands + 50 hundreds ◯ 5 thousands + 6 ten thousands

15 Find the difference between 743,082 and 68,267.

16 Find the first five multiples of 9 that are greater than 90.

17 Circle all prime numbers and cross off all composite numbers.

| 2 5 17 21 48 67 89 |

18 List the common factors of 48 and 84.

19 Find the product.

36 × 58 = ☐

20 7,059 ÷ 8 is _____ with a remainder of _____.

Section C (4 points each)

21 There were 3,683 more liters of water in Tank A than in Tank B. After 1,982 liters of water were drained from Tank B, there were 5,820 liters of water left in Tank B. How many liters of water are in Tank A?

22 A local charity raised $18,064 in January. If it raises this same amount each month, how much money will it raise by the end of June?

23 Charlie made bracelets to sell at the Saturday market. He used 189 red beads and 22 times as many white beads as red beads. How many beads did he use altogether?

24 A candy store sold twice as many chocolates on Saturday as on Friday, and 20 more chocolates on Thursday than on Friday. During the three days, it sold 1,980 chocolates altogether. How many chocolates did it sell on Thursday?

25 Class A collected 5 times as many cans of food as Class B for a food drive. Altogether, they collected 1,056 cans. A parent then donated an additional 260 cans to each class. How many cans of food did Class B collect in total?

BLANK

Name: _____

Date: _____

50 min   Score

60

Test B

Continual Assessment 1

Section A (2 points each)
Circle the correct option: **A**, **B**, **C**, or **D**.

1 In 324,048 the value of the digit in the ten thousands place is _____.

   A 320,000   **B** 20,000

   C 24,048   **D** 4,000

2 ☐ × 8 = 72,000

   A 7,000   **B** 9,000

   C 90,000   **D** 20,000

3 Find the sum of 68,024, 4,392, and 9,062.

   A 81,478   **B** 82,423

   C 80,978   **D** 72,416

4 9,984 − 3,998 = ☐

A 5,986

B 6,086

C 5,842

D 5,982

5 Three lights flash in intervals of 4, 6, and 8 seconds. If they all just flashed together, in how many more seconds will they flash together again?

A 40

B 8

C 16

D 24

6 What is the greatest factor of 64, other than the number itself?

A 16

B 32

C 48

D 8

7 5,999 × 6 = ☐

    A 36,006   **B** 35,999

    C 35,994   **D** 36,994

8 A farm has 80 blueberry bushes in 112 rows. How many blueberry bushes are there altogether?

    A 8,960   **B** 9,860

    C 8,720   **D** 9,220

9 Which expression is equal to 1,200?

    A 9,800 ÷ 8   **B** 8,200 ÷ 7

    C 10,800 ÷ 9   **D** 7,000 ÷ 6

10 Which expression has the same quotient as 6,874 ÷ 7?

    A 4,910 ÷ 5   **B** 3,932 ÷ 4

    C 6,867 ÷ 7   **D** 8,820 ÷ 9

Section B (2 points each)

11 Count on by 2 ten thousands and 1 hundred to complete the number pattern.

| 174,989 | | | | |

12 Mark the approximate location of 951,019 on the number line. Then round 951,019 to the nearest hundred thousand.

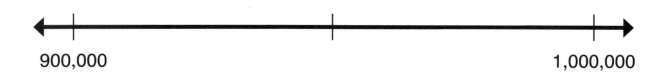

900,000   1,000,000

951,019 is _____ when rounded to the nearest hundred thousand.

13 Write the missing number.

989,746 = 700 + 6 + ☐ + 40 + 80,000

14 Write > or < in the ◯.

732,004 − 100 ◯ 20 tens + 7 hundred thousands + 30 hundreds

15 54,338 − ☐ = 23,482

16 What is the greatest multiple of 9 that is less than 300?

17 List the first five prime numbers and the first five composite numbers.

Prime numbers: _____

Composite numbers: _____

18 Circle the two numbers that have exactly 3 common factors.

24   45   72   2

19 Find the product.

92 × 41 = ☐

20 9,683 ÷ 7 is _____ with a remainder of _____.

Section C (4 points each)

21 A giraffe weighs 9,862 lb less than an elephant and 1,027 lb more than a tiger. The tiger weighs 582 lb. How much does the elephant weigh?

22 A laptop costs $1,189. The math department at a school purchased 5 laptops and the science department purchased 9 laptops. How much more money did the science department spend on laptops than the math department?

23 Over the summer, Sam sold berries at the farmer's market. She received $280 from blueberry sales and 11 times as much money from blackberry sales as from blueberry sales. She received $680 less from strawberry sales than from blackberry sales. How much did she receive altogether?

24 A school ordered 7,930 books. It ordered twice as many math books as history books, 320 more science books than history books, and 40 fewer spanish books than history books. How many spanish books did it order?

25 Two fields had a total of 278 sheep. After 46 lambs were added to North Field, and 14 sheep in South Field were sold, South Field had 26 more sheep than North Field. How many sheep were in each field at first?

BLANK

Name: _____

Date: _____

25 min   Score

30

Test A

Chapter 6 Fractions

Section A (2 points each)
Circle the correct option: **A, B, C,** or **D**.

1 _____ is an equivalent fraction of $\frac{2}{5}$.

    A $\frac{2}{10}$ **B** $\frac{4}{20}$

    C $\frac{6}{25}$ **D** $\frac{20}{50}$

2 The simplest form of $\frac{48}{64}$ is _____.

    A $\frac{7}{8}$ **B** $\frac{3}{4}$

    C $\frac{12}{16}$ **D** $\frac{1}{2}$

3 Which of the following fractions is closest to 1?

A $\frac{10}{11}$ B $\frac{2}{3}$

C $\frac{5}{12}$ D $\frac{7}{8}$

4 What number is indicated by P on the number line?

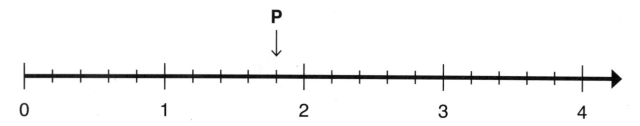

A $\frac{5}{9}$ B $\frac{4}{5}$

C $\frac{9}{5}$ D $\frac{9}{2}$

5 What is 3 ÷ 6 expressed as a fraction in simplest form?

A $\frac{3}{6}$ B $\frac{3}{4}$

C 2   D $\frac{1}{2}$

Section B (2 points each)

6 Express $\frac{18}{63}$ in simplest form.

7 Write >, <, or = in the ◯.

$\frac{1}{2}$ ◯ $\frac{24}{48}$

8 Circle the two fractions indicated by the arrows.

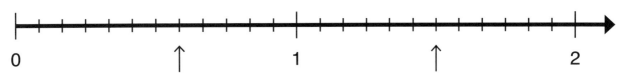

$\frac{7}{10}$ $\frac{7}{12}$ $1\frac{5}{12}$ $1\frac{1}{2}$

9 Put the fractions in order from least to greatest.

$$\frac{8}{13}, \frac{11}{13}, \frac{3}{3}, \frac{3}{13}$$

_____ _____ _____ _____

10 Express $4 + \frac{3}{9}$ as a mixed number in simplest form.

11 What is 66 ÷ 8 expressed as a mixed number or whole number in simplest form? Circle the answer.

$$8 \quad 8\frac{2}{8} \quad 8\frac{1}{3} \quad 8\frac{1}{4}$$

12 Express $2\frac{1}{12}$ as an improper fraction.

13 Circle the fractions that are greater than $\frac{17}{3}$.

$$4\frac{2}{3} \quad \frac{30}{5} \quad \frac{59}{10} \quad 5\frac{1}{6}$$

14 Gavin pours 3 L of juice equally into 8 bottles. How many liters of juice are in each bottle?

15 A white pumpkin weighs $9\frac{2}{5}$ lb. An orange pumpkin weighs $9\frac{3}{15}$ lb. Which pumpkin is lighter?

Name: _____

Date: _____

25 min   Score

30

Test B

Chapter 6 Fractions

Section A (2 points each)
Circle the correct option: **A**, **B**, **C**, or **D**.

1 The simplest form of $\frac{8}{100}$ is _____.

    A $\frac{8}{10}$ **B** $\frac{2}{25}$

    C $\frac{2}{5}$ **D** $\frac{20}{25}$

2 Which of the following is an improper fraction?

    A $1\frac{6}{7}$ **B** $\frac{60}{70}$

    C $\frac{6}{7}$ **D** $\frac{7}{6}$

3 The simplest form of $1\frac{6}{30}$ is _____.

A $1\frac{2}{10}$

B $\frac{36}{30}$

C $1\frac{1}{5}$

D $\frac{16}{30}$

4 Which of the following fractions is closest to 6?

A $\frac{59}{10}$

B $\frac{27}{5}$

C $\frac{63}{10}$

D $\frac{13}{2}$

5 What is 90 ÷ 4 expressed as a mixed number in simplest form?

A $22\frac{2}{4}$

B 22

C $22\frac{1}{2}$

D $20\frac{3}{4}$

Section B (2 points each)

6 Express $\frac{15}{36}$ in simplest form.

7 Write >, <, or = in the ◯.

$\frac{1}{2}$ ◯ $\frac{3}{18}$

8 Express $\frac{7}{28} + 7$ as a mixed number in simplest form.

9 Express $\frac{57}{9}$ as a mixed number in simplest form.

10 Circle the fractions that are equivalent to $2\frac{1}{2}$.

$$\frac{22}{4} \quad \frac{8}{4} \quad \frac{15}{6} \quad \frac{14}{6} \quad \frac{20}{8}$$

11 Label $1\frac{4}{16}$ and $\frac{11}{5}$ on the number line each with arrows.

12 Write the missing number.

$$\frac{3}{9} = \boxed{\frac{}{12}}$$

13 Write the numbers in order from greatest to least.

$$\frac{20}{8}, \ 3\frac{6}{7}, \ 2\frac{1}{3}, \ \frac{28}{7}$$

_____ _____ _____ _____

14 Lisa cut 58 m of ribbon into 5 equal-length pieces. How long is each piece in meters?

15 Emma spent $\frac{3}{4}$ of an hour doing homework, $\frac{7}{12}$ of an hour playing with her neighbors, and $\frac{3}{5}$ of an hour reading. On which activity did Emma spend the longest time?

Name: _____

Date: _____

25 min   Score

30

Test A

Chapter 7 Adding and Subtracting Fractions

Section A (2 points each)
Circle the correct option: **A**, **B**, **C**, or **D**.

1 $\frac{1}{4} + \frac{1}{8} + \frac{1}{2} =$ _____

   A $\frac{3}{8}$ **B** $\frac{3}{4}$

   C $\frac{7}{8}$ **D** $\frac{5}{8}$

2 $\frac{5}{8} - \frac{11}{40} =$ _____

   A $\frac{7}{20}$ **B** $\frac{7}{40}$

   C $\frac{13}{40}$ **D** $\frac{9}{10}$

Chapter 7 Test A   97

3 $3\frac{2}{7} + \frac{9}{14} =$ _____

    A $4\frac{13}{14}$ **B** $3\frac{7}{13}$

    C $3\frac{11}{14}$ **D** $3\frac{13}{14}$

4 $9 - 2\frac{7}{10} =$ _____

    A $6\frac{7}{10}$ **B** $11\frac{7}{10}$

    C $6\frac{3}{10}$ **D** $2\frac{3}{10}$

5 $8\frac{1}{3} - 3\frac{1}{9} =$ _____

    A $11\frac{1}{9}$ **B** $5\frac{2}{9}$

    C $5\frac{2}{3}$ **D** $5\frac{1}{9}$

Section B (2 points each)

Express your answers in proper fraction, whole number, or mixed numbers in simplest form.

6 $\dfrac{4}{15} + \dfrac{7}{15} + \dfrac{14}{15} = $ _____

7 $\dfrac{13}{9} + \dfrac{5}{18} = $ _____

8 $\dfrac{10}{21} - \dfrac{2}{7} = $ _____

9 $1\frac{3}{4} - \frac{5}{12} =$ _____

10 $7\frac{1}{2} + 7\frac{3}{4} =$ _____

11 Chapa ran $\frac{1}{2}$ mile yesterday. She ran $\frac{11}{18}$ mile today. How many miles did she run altogether in the two days?

Section C (4 points each)

12 John had three bags of beans that weighed $1\frac{2}{5}$ kg, $2\frac{2}{15}$ kg, and $3\frac{1}{5}$ kg. He poured all the beans from the three bags into an empty container. How many kilograms of beans are in the container?

13 Amy had $5\frac{3}{4}$ m of ribbon. She used $2\frac{1}{8}$ m to tie a present and $1\frac{1}{2}$ m to make a bow. How many meters of ribbon does she have left?

Name: _____

Date: _____

25 min   Score

30

Test B

Chapter 7 Adding and Subtracting Fractions

Section A (2 points each)
Circle the correct option: **A**, **B**, **C**, or **D**.

1 $\frac{2}{25} + \frac{1}{50} + \frac{3}{5} =$ _____

  A $\frac{15}{25}$
  B $\frac{6}{50}$
  C $\frac{7}{10}$
  D $\frac{33}{50}$

2 $\frac{13}{14} - \frac{3}{7} =$ _____

  A $1\frac{3}{4}$
  B $\frac{1}{2}$
  C $\frac{1}{14}$
  D $\frac{10}{14}$

Chapter 7 Test B   103

3 $\frac{4}{9} + 4\frac{7}{18} =$ _____

    A $4\frac{1}{18}$ **B** $4\frac{11}{18}$

    C $4\frac{1}{2}$ **D** $4\frac{5}{6}$

4 $5\frac{3}{4} + 1\frac{3}{16} + 2\frac{1}{8} =$ _____

    A $8\frac{1}{16}$ **B** $9\frac{1}{16}$

    C $1\frac{1}{16}$ **D** $9\frac{3}{16}$

5 $3\frac{7}{12} - 2\frac{5}{6} =$ _____

    A $\frac{3}{4}$ **B** $1\frac{1}{6}$

    C $\frac{1}{2}$ **D** $\frac{5}{12}$

Section B (2 points each)

Express your answers as a proper fraction, whole number, or mixed number in simplest form.

6 $5\frac{1}{5} - 1\frac{2}{5} = $ _____

7 $\frac{1}{2} + \frac{2}{4} + 1\frac{3}{8} = $ _____

8 $6\frac{1}{16} - \frac{7}{8} = $ _____

9 $\frac{10}{9} - \frac{1}{3} =$ _____

10 Holly bought $1\frac{1}{5}$ lb of yellow potatoes and $2\frac{7}{10}$ lb of red potatoes. How many pounds of potatoes did she buy altogether?

11 Justin spent $\frac{5}{8}$ of an hour writing in his scrapbook. He spent $\frac{1}{2}$ an hour drawing in his scrapbook. How much longer did he spend writing than drawing in his scrapbook? Express your answer as a fraction of an hour.

Section C (4 points each)

12 Hannah had a full can of paint. She used $\frac{2}{3}$ of it to paint a rocking chair and $\frac{2}{9}$ of it to paint a bird feeder. What fraction of the can of paint does she have left?

13 Eli bought $3\frac{1}{6}$ m of ribbon. Kona bought $\frac{5}{12}$ m less ribbon than Eli. How many meters of ribbon did they buy altogether?

Name: _____

Date: _____

25 min   Score

30

Test A

Chapter 8 Multiplying a Fraction and a Whole Number

Section A (2 points each)
Circle the correct option: **A**, **B**, **C**, or **D**.

1 $\frac{1}{7} + \frac{1}{7} + \frac{1}{7} + \frac{1}{7} + \frac{1}{7} + \frac{1}{7} =$ _____

A $\frac{3}{8}$

B $6 \times \frac{1}{7}$

C $\frac{7}{8}$

D $\frac{5}{8}$

2 What fraction of the set is shaded?

A $\frac{3}{4}$

B $\frac{3}{10}$

C $\frac{1}{4}$

D $\frac{1}{3}$

Chapter 8 Test A   109

3 $12 \times \frac{1}{8} =$ _____

    A $1\frac{1}{2}$ **B** $1\frac{1}{4}$

    C $1\frac{3}{8}$ **D** $12\frac{1}{8}$

4 $\frac{1}{5} \times 100 =$ _____

    A 50   **B** 20

    C 2   **D** 25

5 What is $\frac{1}{3}$ of $90?

    A $10   **B** $90

    C $45   **D** $30

Section B (2 points each)

Express answers in simplest form.

6 $\frac{2}{3} \times 5 = $ _____

7 $\frac{5}{6} \times 9 = $ _____

8 $12 \times \frac{3}{4} = $ _____

9 $\frac{4}{7} \times 21 = $ _____

10 There are 24 apples in a basket. $\frac{5}{6}$ of them are red apples. How many red apples are in the basket?

11 There are 30 red and green apples in a basket. 16 of them are red. What fraction apples in the basket are green?

Section C (4 points each)

12 Jenna gave $\frac{2}{5}$ of her stickers to her brother. She gave her brother 18 stickers. How many stickers did she have at first?

13 Landon bought some peaches. He used $\frac{1}{3}$ of the peaches to make pies and the rest to make jam. He used 16 peaches to make jam. How many peaches did he use to make pies?

Name: _____

Date: _____

25 min   Score

30

Test B

Chapter 8 Multiplying a Fraction and a Whole Number

Section A (2 points each)

Circle the correct option: **A**, **B**, **C**, or **D**.

1 $\frac{1}{3} \times 4$ is not equal to _____.

    A $\frac{4}{3}$ **B** $\frac{1}{3} + \frac{1}{3} + \frac{1}{3} + \frac{1}{3}$

    C $\frac{1}{4} \times 3$ **D** $1\frac{1}{3}$

2 $6 \times \frac{7}{10} =$ _____

    A $\frac{3}{5}$ **B** $\frac{13}{10}$

    C $4\frac{1}{6}$ **D** $4\frac{1}{5}$

3 $\frac{3}{8} \times 14 =$ _____

    A $8\frac{1}{16}$ **B** $3\frac{4}{7}$

    C $5\frac{1}{4}$ **D** $9\frac{3}{16}$

4 What is $\frac{3}{5}$ of $100?

    A $30   **B** $60

    C $45   **D** $180

5 What fraction of a 120-page book is 40 pages?

    A $\frac{1}{3}$ **B** $\frac{1}{4}$

    C $\frac{1}{5}$ **D** $\frac{2}{3}$

Section B (2 points each)

Express answers in simplest form.

6 $8 \times \frac{5}{7} =$ _____

7 $6 \times \frac{1}{8} =$ _____

8 What is $\frac{3}{4}$ of 50?

9 A cook bought 20 yellow potatoes, 15 red potatoes, and 5 purple potatoes. What fraction of the potatoes are purple potatoes?

10 Nate bought 7 pizzas to share equally among 35 people. What fraction of a pizza did each person get?

11 In a class of 24 children, $\frac{7}{12}$ of them are girls. How many boys are in the class?

Section C (4 points each)

12 There are 160 balls in a ball pit. $\frac{3}{20}$ of them are red. There are twice as many blue balls as red balls in the pit. The rest of the balls are green. How many red and blue balls are in the ball pit altogether?

13 Hannah's mom gave her some money to spend at a fair. Hannah spent $4 on snacks. She then spent $\frac{4}{7}$ of the rest of her money on rides. She has $6 left. How much money did her mom give her?

Name: _____

Date: _____

25 min Score

30

Test A

Chapter 9 Line Graphs and Line Plots

Section A (2 points each)
Circle the correct option: **A, B, C,** or **D**.

1 This table shows the number of birthday cakes sold by a bakery last week. Which graph is an accurate representation of the data?

Mon	Tue	Wed	Thu	Fri	Sat	Sun
5	15	35	15	60	60	30

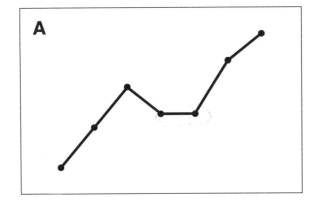

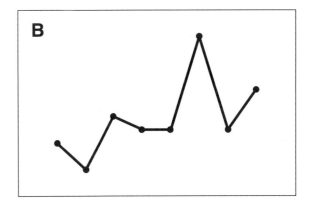

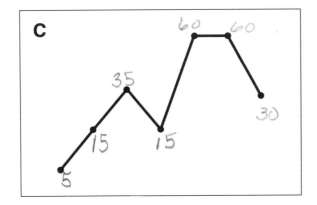

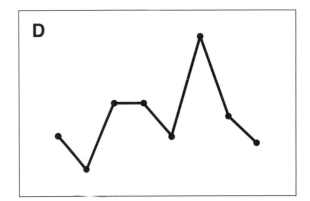

The line plot shows the number of pairs of women's shoes by size sold by a store yesterday. Use the line plot to answer questions 2–5.

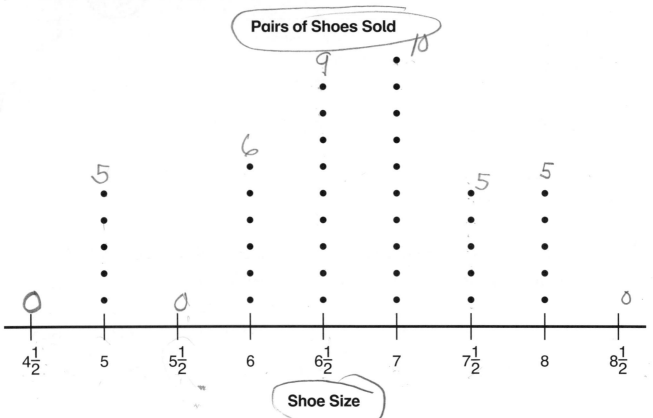

2 How many pairs of shoes were sold?

A 30

B 40

C 36

D 9

3 What was the most common shoe size sold?

A $6\frac{1}{2}$

B $5\frac{1}{2}$

C 8

D 7

4 How many pairs of $5\frac{1}{2}$ shoes were sold?

A 0

B 4

C 10

D 5

5 How many more pairs of $6\frac{1}{2}$ shoes than $7\frac{1}{2}$ shoes were sold?

A 16

B 2

C 4

D 6

Section B (2 points each)

The table shows the number of days it rained each month from September through April in Seattle. A graph of the same data is partially completed. Use the data to answer questions 6–10.

Month	Sep	Oct	Nov	Dec	Jan	Feb	Mar	Apr
Number of Days it Rained	8	14	17	19	19	15	16	13

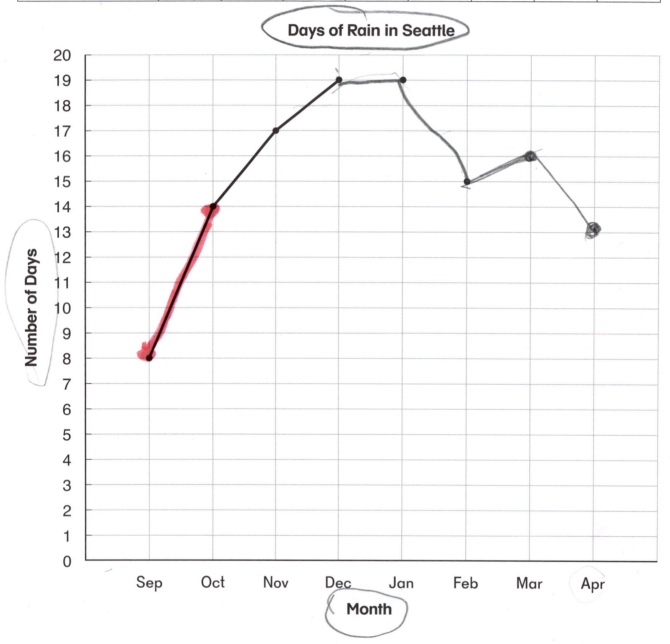

124   Chapter 9 Test A

6 Complete the graph.

7 Which month had the least number of days of rain?

September

8 Which months had more than 18 rainy days per month?

Dec
Jan

9 Between which two months was the sharpest increase in number of rainy days?

10 Which month had 4 fewer rainy days than November?

Chapter 9 Test A

The line plot shows the number of runs scored in each game by a baseball team last month. Use the line plot to answer questions 11–15.

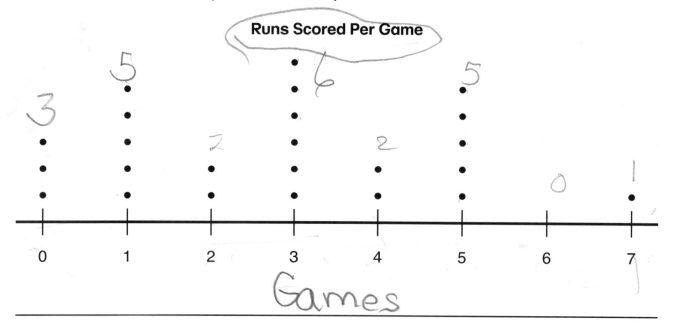

11 What does the horizontal axis represent?

12 How many games are recorded on the line plot?

13 What is the difference between the greatest and least number of runs scored?

14 In how many games did the team score more than 3 runs?

15 What fraction of the total games did they score no runs?

Name: _____

Date: _____

25 min   Score

30

Test B

Chapter 9 Line Graphs and Line Plots

Section A (2 points each)
Circle the correct option: **A**, **B**, **C**, or **D**.

Alex made a survey on the number of pets owned by his friends and recorded the data on a line plot.

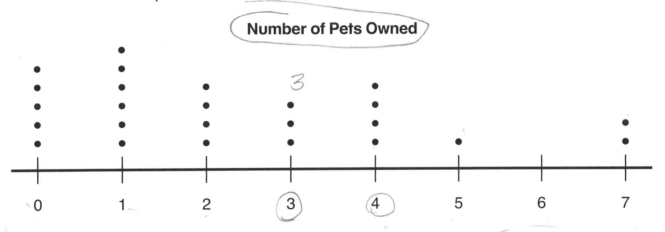

1. How many friends did Alex survey?

   A 28   **B** 23

   C 25   **D** 7

2. How many of Alex's friends have more than 2 pets?

   A 10   **B** 27

   C 15   **D** 14

The graph shows the amount of money saved by Ella monthly from January through June. Use the graph to answer questions 3–5.

3. How many dollars does each increment between tick marks represent?

A 1

B 2

C 5

D 10

8 Between which two days was the sharpest decrease in sales of ice cream cones?

9 On how many days did the shop sell more than 160 ice cream cones per day?

10 Each ice cream cone cost $3. How much more money did the shop receive from selling ice cream cones on ___day than on Monday?

Chapter 9 Test B

The table shows the weight in pounds of bags of grapes a store sold. Use the table to answer questions 11–15.

$3\frac{1}{4}$	$2\frac{3}{4}$	$3\frac{1}{2}$	$3\frac{1}{2}$	$3\frac{3}{4}$	$3\frac{1}{4}$	$3\frac{3}{4}$	$3\frac{1}{2}$	$3\frac{1}{2}$
4	$3\frac{1}{2}$	$3\frac{1}{2}$	$3\frac{1}{4}$	4	$3\frac{1}{2}$	$2\frac{3}{4}$	$3\frac{3}{4}$	$2\frac{1}{2}$
4	$3\frac{1}{2}$	$2\frac{3}{4}$	$3\frac{3}{4}$	4	$3\frac{1}{4}$	$3\frac{3}{4}$	$3\frac{1}{2}$	

11 Create a line plot.

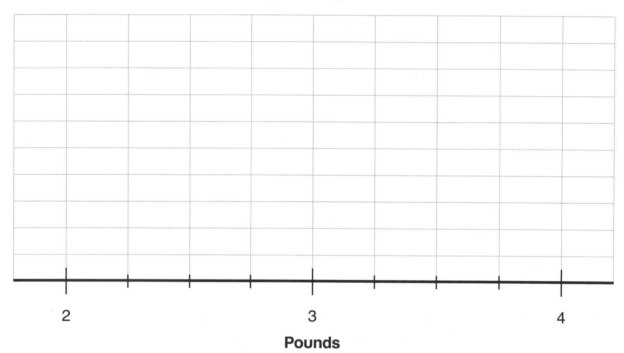

12 How many bags of grapes were sold altogether?

13 How many bags weighed more than 4 lb?

14 What was the difference in weight between the heaviest and the lightest bags of grapes?

15 What fraction of the total number of bags sold weighed $3\frac{1}{4}$ lb?

BLANK

Name: _____

Date: _____

50 min   Score

60

Test A

Continual Assessment 2

Section A (2 points each)
Circle the correct option: **A**, **B**, **C**, or **D**.

1 What is the sum of 5 tens, 80 ten thousands, and 4 thousands?

   A 8,450   **B** 84,050

   C 840,050   **D** 804,050

2 Which one is equal to 76,541 + 6,716 + 2,345? Use estimation.

   A 89,678   **B** 85,602

   C 80,386   **D** 82,302

3 22,621 − ▢ = 1,378

   A 20,486   **B** 21,243

   C 2,783   **D** 24,586

4 Which of the following numbers is a prime number?

    A 2   B 4

    C 9   D 15

5 18,421 × 8 = ☐

    A 143,551   B 147,368

    C 118,546   D 98,321

6 Which expression has the same quotient as 6,216 ÷ 7?

    A 4,440 ÷ 5   B 7,384 ÷ 6

    C 6,230 ÷ 8   D 8,172 ÷ 9

7 48 ÷ 7 is _____ when expressed as a mixed number.

    A $6\frac{6}{7}$ B $\frac{48}{7}$

    C $6\frac{5}{7}$ D $5\frac{6}{7}$

8 $2\frac{2}{3} + \frac{7}{9} = $ _____

    A $3\frac{1}{3}$ **B** $3\frac{4}{9}$

    C $2\frac{8}{9}$ **D** $3\frac{2}{9}$

9 $12 \times \frac{4}{7} = $ _____

    A $6\frac{5}{12}$ **B** $5\frac{4}{7}$

    C $6\frac{6}{7}$ **D** $7\frac{1}{7}$

10 The table shows the number of books checked out at the library last week.

Day	Mon	Tues	Wed	Thur	Fri	Sat	Sun
Number of Books	85	92	71	88	41	101	95

Which graph represents the data?

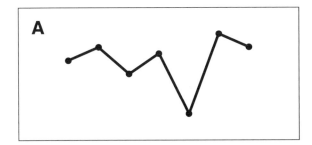
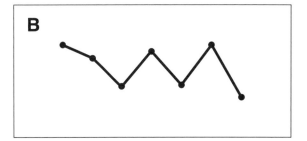
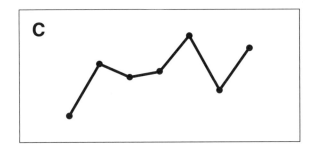
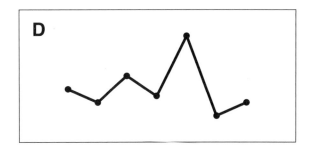

Section B (2 points each)

11 Circle the numbers that are 52,000 when rounded to the nearest thousand.

| 52,799 | 51,801 | 52,499 | 52,500 | 51,499 |

12 Check (✓) the statements that are true.

108 is a multiple of 6.	
78 is a multiple of 9.	
8 is a factor of 64.	

13 Circle the numbers that have 3 and 8 as common factors.

| 18   24   45   72   95 |

14 88 × 52 = ☐

15 Arrange the following numbers in order from least to greatest. Write each number in simplest form.

$$\frac{8}{9}, \frac{2}{4}, \frac{5}{8}, \frac{2}{6}$$

_____ _____ _____ _____

16 Label each arrow with an improper fraction above the number line and a mixed number below the number line. Use simplest form.

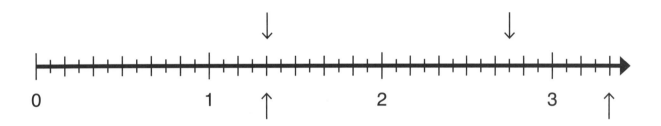

17 Mei brought $8\frac{1}{3}$ L of water camping. On the first day, she drank $3\frac{5}{6}$ L of water. How many liters of water does she have left?

The table shows the number of books students read over the summer. Use the table to answer questions 18–20.

3	5	7	1	0	6	5	5	4	2	1	2	5	4	4
4	3	0	2	6	7	9	3	4	8	1	3	5	5	3

18 Complete the line plot using the data in the table above.

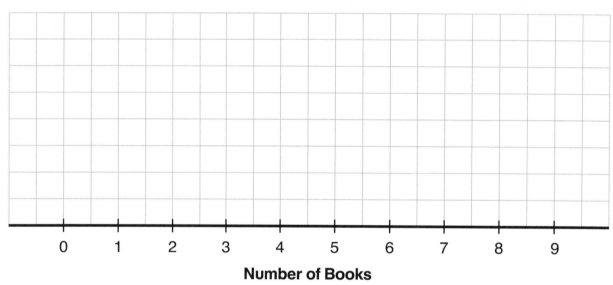

19 What was the most common number of books read?

20 What fraction of the students read 3 books?

142   Continual Assessment 2 Test A

Section C (4 points each)

21 1,512 people attended a festival on Friday, 1,718 more people attended the festival on Saturday than on Friday, and 168 fewer people attended the festival on Sunday than on Saturday. How many people attended the festival on Sunday?

22 Alex bought 17 reams of paper. Each ream contains 350 sheets of paper. She used 780 sheets of paper. How many sheets of paper are left?

A golf course spent $13,880 on 7 new golf carts. Two of the golf carts were larger and were $815 more expensive than each of the other 5 golf carts. How much did one larger golf cart cost?

A baker made 372 chocolate chip cookies and sugar cookies altogether in the morning. After the first customer of the day purchased 24 sugar cookies, there were 52 more chocolate chip cookies than sugar cookies. How many of each type of cookie was there at first?

25 $\frac{5}{9}$ of the trees in an orchard are apple trees. The remaining 60 trees are orange trees. How many apple trees are in the orchard?

Name: _____

Date: _____

50 min   Score

60

Test B

Continual Assessment 2

Section A (2 points each)
Circle the correct option: **A**, **B**, **C**, or **D**.

1 Which one is the sum of 30 tens, 83 ten thousands, and 20 thousands?

  A 83,230   **B** 85,300

  C 832,030   **D** 850,300

2 Which one is equal to 476,541 + 216,716 + 3,347? Use estimation.

  A 671,284   **B** 714,254

  C 696,604   **D** 627,546

3 378,621 − 5,283 = 402,813 − ☐

  A 32,451   **B** 29,475

  C 30,821   **D** 28,813

Continual Assessment 2 Test B   147

4 Which of the following numbers is a prime number?

　　A 87　　　　　　　　　　B 69

　　C 43　　　　　　　　　　D 77

5 18,421 × 8 = ☐

　　A 36,842 × 4　　　　　　B 25,471 × 6

　　C 31,842 × 5　　　　　　D 22,891 × 7

6 Which one has 9 as a factor?

　　A 308　　　　　　　　　 B 679

　　C 600　　　　　　　　　 D 252

7 Add 48 ÷ 9 to $\frac{4}{9}$. What is the answer?

　　A $5\frac{7}{9}$　　　　　　　　　　B $\frac{48}{9}$

　　C $6\frac{1}{9}$　　　　　　　　　　D $5\frac{1}{3}$

8 $3\frac{1}{3} + \frac{7}{9} + \frac{4}{9} =$ _____

   A $4\frac{1}{3}$
   B $3\frac{5}{9}$
   C $3\frac{8}{9}$
   D $4\frac{5}{9}$

9 Which expression is equivalent to $\frac{1}{3}$ of 12?

   A $6 \times \frac{2}{3}$
   B $12 \times \frac{2}{3}$
   C $3 \times \frac{1}{3}$
   D $\frac{1}{3} \times 6$

10 The table shows the number of ice cream cones sold at an ice cream stand last week.

Day	Mon	Tues	Wed	Thur	Fri	Sat	Sun
Number of Cones	92	55	47	111	67	118	131

Which graph does not represent the data?

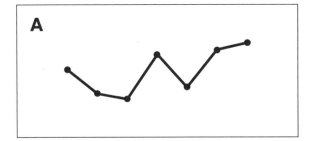

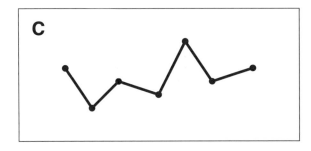

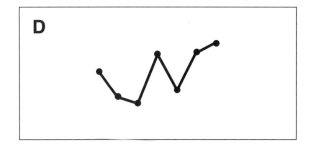

Section B (2 points each)

11 Circle the numbers that are 100,000 when rounded to its highest place value.

| 52,799 | 93,621 | 95,000 | 123,500 | 149,889 |

12 Check (✓) the statements that are true.

All even numbers have at least one prime number as a factor.	
1, 4, and 9 have an odd number of factors.	
64 is a factor of 8.	

13 Circle the numbers that have 4, 6, and 12 as common factors.

| 10 | 18 | 24 | 45 | 72 | 88 | 96 |

14 Arrange the following numbers in order from least to greatest.

$\frac{13}{6}, \frac{16}{5}, \frac{13}{8}, 1\frac{1}{2}, 2\frac{3}{4}, \frac{9}{7}$

_____ _____ _____ _____ _____ _____

15 Circle the expression with the greatest product. Cross out the expression with the least product.

> 32 × 81   51 × 27   18 × 63   78 × 36

16 Label each arrow with a fraction above the number line and a mixed number below the number line. Use simplest form.

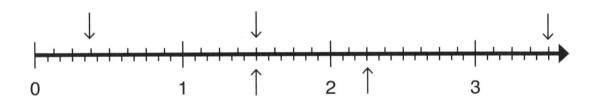

17 A store owner had $9\frac{3}{4}$ kg of apples. On Tuesday, he sold $3\frac{1}{2}$ kg of apples and on Wednesday he sold $2\frac{1}{4}$ kg of apples. How many kilograms of apples does he have left?

The table shows the heights, measured to the nearest half centimeters, of students' bean plants in a fourth grade class. Use the table to answer questions 18–20.

$17\frac{1}{2}$	17	$15\frac{1}{2}$	$14\frac{1}{2}$	16	17	14	$17\frac{1}{2}$	18	$15\frac{1}{2}$	$16\frac{1}{2}$	17	17	18	$16\frac{1}{2}$
$16\frac{1}{2}$	15	$15\frac{1}{2}$	19	15	$14\frac{1}{2}$	$17\frac{1}{2}$	$18\frac{1}{2}$	17	16	17	$16\frac{1}{2}$	19	$15\frac{1}{2}$	16

18 Complete the line plot using the data in the table above.

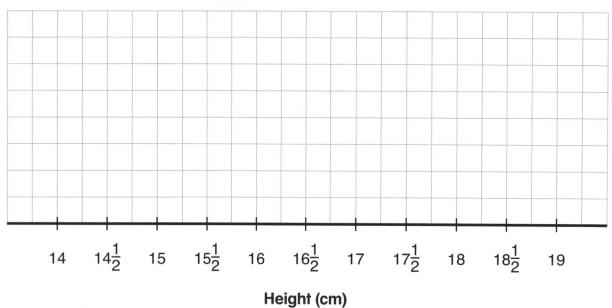

Heights of Students' Bean Plants

Height (cm)

19 How many bean plants are shorter than 16 cm?

20 What fraction of the bean plants are taller than 17 cm?

Section C (4 points each)

21 There are three hiking trails that hikers can take to get to a lake. Trail A is 2,245 m long, Trail B is 425 m longer than Trail A, and Trail C is 187 m shorter than Trail B. How much longer is Trail C than Trail A?

22 There are 24 bottles of juice. Each bottle contains 750 mL of juice. Francesca and her friends drank 3 full bottles and 300 mL of a fourth bottle. How many mL of juice are left?

23 An athletic club had $14,200 to spend on equipment. It purchased 5 bikes and 3 kayaks. The kayaks were $200 more than twice the price of the bikes. There is $400 left over. How much money did each kayak cost?

24 An ice cream shop had 650 pints of chocolate ice cream and 170 pints of vanilla ice cream. A customer bought the same number of pints of each flavor for a company party. The ice cream shop now has 5 times as many pints of chocolate as vanilla ice cream. How many pints did the customer buy?

25 On a three-day backpacking trip, Jacob hiked 3 miles the first day, 7 miles the second day, and $\frac{1}{3}$ of the total distance on the third day. How many miles did he hike altogether?

Answer Key

Detailed solutions given are suggestions, and do not include all possible methods of arriving at the correct answer. Accept all reasonable solutions by students.

Test A

Chapter 1 Numbers to One Million

1 A

2 C

3 B

4 C

5 D

6 69,016

7 58,016

8 fifty-nine thousand, sixteen

9 50,000 + 9,000 + 10 + 6

10 277,352

11 300,000

12 450,000

13 136,263 | 137,264 | **138,265**
139,266 | 140,267 | **141,268**

14 255,900 | 525,100 | 550,001
801,550 | 900,000

15 $4,000

Test B

Chapter 1 Numbers to One Million

1 C

2 D

3 C

4 C

5 A

6 410,682

7 300,682

8 thousands, 0

9 300,000 + 10,000 + 600 + 80 + 2

10 930,925

11 900,000

12 950,000

13 103,589

14 326,211, 352,058, 357,249, 633,621

or

C, B, A, D

15 2,000 × 6 = 12,000
12,000 + 3,000 = 15,000
$15,000

Test A

Chapter 2 Addition and Subtraction

1 B

2 D

3 A

4 C

5 B

6 38,753 ≈ 40,000
51,335 ≈ 50,000
40,000 + 50,000 = 90,000
90,088

7 741,054 ≈ 750,000
46,281 ≈ 50,000
750,000 − 50,000 = 700,000
4 − 1 = 3, so last digit must be 3.
694,773

8 65,300 − 49
    /   \
 65,250   50
50 − 49 = 1
65,250 + 1 = 65,251
65,251

9 281,948

10 7,036

11 4,911 eggs

12
Scooter: 1,250
Helmet: 1,125
?

1,250 − 1,125 = 125
1,250 + 125 = 1,375
$1,375

13
Men: 5,500
Women: 12,000
Children: 100, ?

5,500 + 5,500 + 100 = 11,100
12,000 − 11,100 = 900
900 children

Test B

Chapter 2 Addition and Subtraction

1 B

2 A

3 C

4 B

5 A

6 6,831 ≈ 7,000
71,582 ≈ 70,000
1,984 ≈ 2,000
7,000 + 70,000 + 2,000
= 79,000
1 + 2 + 4 = 7, so last digit must be 7.
80,397

7 320,176 ≈ 330,000
28,524 ≈ 30,000
330,000 − 30,000 = 300,000
Answer cannot be greater than 320,176.
6 − 4 = 2, so last digit must be 2.
291,652

8 59,735 − 28
　　／　＼
　59,705　30

30 − 28 = 2
59,705 + 2 = 59,707
59,707

9 367,074

10 202,132

11 $226

12
Friday: 8,152
Saturday: 1,250
Sunday: ?
899

Saturday: 8,152 + 1,250 = 9,402
Sunday: 9,402 − 899 = 8,503
9,402 + 8,503 = 17,905
17,905 people

Test B

Chapter 2 Addition and Subtraction

13 Before — Warehouse 2,856; Store 2,002

After — Warehouse; Store 1,449

Store before: 2,856 − 2,002
  / \
 2,000 2

2,856 − 2,000 = 856
856 − 2 = 854
1,449 − 854 = 595
595 shirts

Test A

Chapter 3 Multiples and Factors

1 B

2 D

3 C

4 C

5 A

6 | 7 (16) 22 28 (32) (40) 44 |

7 6, 12, 18, 24, 30

8 Multiples of 5: 5, 10, 15, 20, 25, 30...
30 is also a multiple of 2 and 3.
30 and 60

9 | (2) 5 (6) 8 (9) 10 (12) |

10 | 30 (17) (5) 27 14 (11) |

11 | 2 (8) 10 (16) (40) |

12 Factors of 18: 1, 2, 3, 6, 9, <u>18</u>
Factors of 36: 1, 2, 3, 4, 6, 9, 12, <u>18</u>, 36
18

13 First four multiples of 7: 7, 14, 21, 28
7 + 14 + 21 + 28 = 70
70

14 Multiples of 5: 5, 10, 15, 20, 25, 30, 35, <u>40</u>, 45...
Multiples of 8: 8, 16, 24, 32, <u>40</u>, 48...
9:40 a.m.

15 Factors of 24: 1, 2, 3, 4, <u>6</u>, 8, 12, 24
Factors of 30: 1, 2, 3, 5, <u>6</u>, 10, 15, 30
6 boxes

Test B

Chapter 3 Multiples and Factors

1 C

2 B

3 A

4 D

5 C

6 9, 18, 27, 36, 45, 54, 63, 72, 81, 90

7 Multiples of 8: 8, 16, 24, 32, 40, 48, 56…
24 and 48 are also multiples of 4 and 6.
24, 48

8 1, 2, 3, 6, 11, 22, 33, 66

9
A prime number is always an odd number.	
A composite number has more than 2 factors.	✓
A multiple of an even number is always even.	✓

10 11, 13, 17, 19

11 8

12 Factors of 22: 1, 2, 11, 22
22

13 Factors of 36: 1, 2, 3, 4…
Multiples of 6: 6, 12, 18, 24, 30, 36
36

14 Multiples of 8: 8, 16, 24, 32, 40, 48…
24 is also a multiple of 4 and 6.
7:24 a.m.

15 Factors of 60: 1, 2, 3, 4, 5, 6, 10, <u>12</u>, 15, 20, 30, 60
Factors of 72: 1, 2, 3, 4, 6, 8, 9, <u>12</u>, 18, 24, 36, 72
Factors of 96: 1, 2, 3, 4, 6, 8, <u>12</u>, 16, 24, 32, 48, 96
12 bags

Test A

Chapter 4 Multiplication

1 B

2 C

3 A

4 C

5 D

6 4,050

7 2,200

8 2,300

9 2,200 is closer to 2,214 than 2,300 is.
Sofia's estimate

10 30,000 × 4 = 120,000 so answer has to be greater than 120,00.
4 × 4 = 16 so the digit in the ones place has to be 6.
130,696

11 3,186

12 192

13 40 × 70 = 2,800
700 × 40 = 28,000
<

14 $1,878

15 Skis: 281
Snow mobile: ?
281 × 9 = 2,529
$2,529

16 Yellow: 920
White
Red
?
920 × 8 = 7,360
7,360 white and red roses

17 1,296 + 3,888 = 5,184
5,184 × 15 = 77,760
77,760 croissants

Test B

Chapter 4 Multiplication

1 C

2 D

3 A

4 C

5 B

6 5,000 × 8 = 40,000
300 × 8 = 2,400
40,000 + 2,400 = 42,400
42,400

7 65,000

8 64,000

9 Dion's estimate because 64,000 is closer to 64,489 than 65,000 is.

10 300 × 80 = 24,000 so answer has to be more than 24,000. 9 × 1 = 9 so the last digit has to be 9. 19 × 1 is greater than 9 so answer has to be greater than 24,009.
25,839

11 88,710

12 2,046

13 5 × 7,000 = 35,000
1,000 × 15 = 15,000
>

14 32,000 × 2 = 64,000
9,000 × 4 = 36,000
11,000 × 3 = 33,000
6,000 × 9 = 54,000
10,999 × 3, 9,039 × 4, 5,500 × 9, 32,185 × 2

Chapter 4 Multiplication

15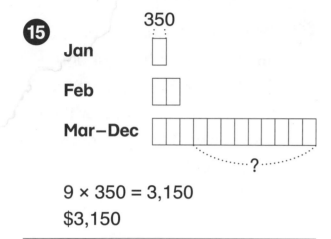

9 × 350 = 3,150

$3,150

16 116 × 6 = 696

696 × 10 = 6,960

6,960 chocolate bars

17 918 × 55 = 50,490

354 × 36 = 12,744

50,490 + 12,744 = 63,234

$63,234

Test A

Chapter 5 Division

1 A

2 C

3 B

4 D

5 B

6 1,000 ÷ 5 = 200; 50 ÷ 5 = 10
200 − 10 = 190
190

7 141, 3

8 848 ÷ 4

9 6,400 ÷ 8 = 800
3,000 ÷ 3 = 1,000
<

10 2,470, 1

11 592, 4

12 200

13 78 boxes

14 $319

15 2,176 ÷ 8 = 272
272 × 7 = 1,904
1,904 cookies

16 First / Second / Third — 504
600 − 96 = 504
504 ÷ 7 = 72
72 counters

17 Beth 80 / Danny / Katie 20 — 1,025
3 units ⟶ 1,025 − 80 − 60 = 885
1 unit ⟶ 885 ÷ 3 = 295
$295

Test B

Chapter 5 Division

1 D

2 C

3 C

4 B

5 A

6 12,000 ÷ 6 = 2,000
3,000 ÷ 6 = 500
2,000 + 500 = 2,500
2,500

7 5,600

8 6,300

9 Dion's because 5,600 is closer to 5,901 than 6,300 is.

10 41, 5

11 1,645, 4

12 <

13

Scooter 1 [ ]
Scooter 2 [ ][ ]   3,090

3 units ⟶ 3,090
1 unit ⟶ 3,090 ÷ 3 = 1,030
$1,030

14 115 + 10 = 125
125 ÷ 5 = 25
25 students and teachers

Test B
Chapter 5 Division

15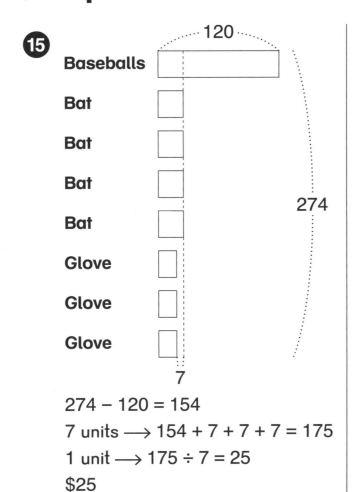

274 − 120 = 154
7 units ⟶ 154 + 7 + 7 + 7 = 175
1 unit ⟶ 175 ÷ 7 = 25
$25

16 8,181 ÷ 9 = 909
909 ÷ 3 = 303
Or
8,181 ÷ 3 = 2,727
2,727 ÷ 9 = 303
303 pieces of candy

17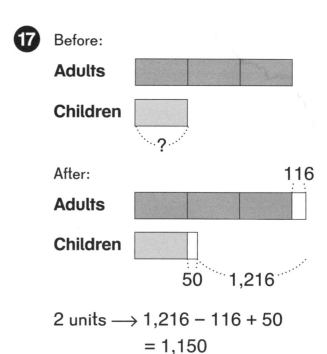

2 units ⟶ 1,216 − 116 + 50
       = 1,150
1 unit ⟶ 1,150 ÷ 2 = 575
575 children

Test A

Continual Assessment 1

1 B

2 A

3 A

4 D

5 B

6 B

7 A

8 C

9 D

10 D

11 91,489   101,489   111,489
121,489   131,489

12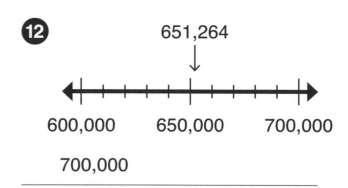
700,000

13 53,625

14 600,000 + 5,000 = 605,000
5,000 + 60,000 = 65,000
>

15 674,815

16 99, 108, 117, 126, 135

17 (2)(5)(17)(~~21~~)(~~48~~)(67)(89)

Test A

Continual Assessment 1

18 Factors of 48: 1, 2, 3, 4, 6, 8, 12, 16, 24, 48
Factors of 84: 1, 2, 3, 4, 6, 7, 12, 14, 21, 28, 42, 84
1, 2, 3, 4, 6, 12

19 2,088

20 882, 3

21 Before:
Tank A — 3,683
Tank B

After: ?
Tank A
Tank B
5,820   1,982

5,820 + 1,982 + 3,683 = 11,485
11,485 liters

22 $108,384

23 White beads: 22 × 189 = 4,158
4,158 + 189 = 4,347
4,347 beads

24
Thur — 20
Fri
Sat
1,980

4 units ⟶ 1,980 − 20 = 1,960
1 unit ⟶ 1,960 ÷ 4 = 490
490 + 20 = 510

510 chocolates

25 Before:
Class A
Class B
1,056

After:
Class A — 260
Class B — 260
?

6 units ⟶ 1,056
1 unit ⟶ 1,056 ÷ 6 = 176
176 + 260 = 436
436 cans

Test B

Continual Assessment 1

1 B

2 B

3 A

4 A

5 D

6 B

7 C

8 A

9 C

10 A

11 174,989 | 195,089 | 215,189 | 235,289 | 255,389

12
951,019 on a number line between 900,000 and 1,000,000

1,000,000

13 909,000

14 732,004 − 100 = 731,904
200 + 700,000 + 3,000
= 703,200
>

15 30,856

16 297

17 Prime numbers: 2, 3, 5, 7, 11
Composite numbers: 4, 6, 8, 9, 10

Test B

Continual Assessment 1

18 Factors of 24: 1, 2, 3, 4, 6, 8, 12, 24
Factors of 45: 1, 3, 5, 9, 15, 45
Factors of 72: 1, 2, 3, 4, 6, 8, 9, 12, 18, 24, 36, 72
Factors of 2: 1, 2

19 3,772

20 1,383, 2

21 582 + 1,027 = 1,609
1,609 + 9,862 = 11,471
11,471 pounds

22 9 − 5 = 4
4 × 1,189 = 4,756
$4,756

23 11 × 280 = 3,080
3,080 − 680 = 2,400
280 + 3,080 + 2,400 = 5,760
$5,760

24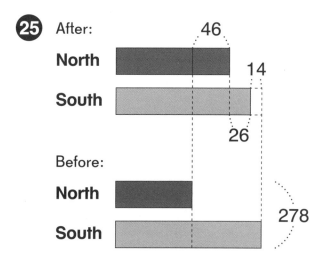

5 units ⟶ 7,930 − 320 + 40
          = 7,650
1 unit ⟶ 7,650 ÷ 5 = 1,530
1,530 − 40 = 1,490
1,490 spanish books

25 [bar model diagram: After — North 46, South 14, difference 26; Before — North, South 278]

2 units ⟶ 278 − 46 − 26 − 14
          = 192
1 unit ⟶ 192 ÷ 2 = 96
96 + 46 + 26 + 14 = 182
North Field: 96 sheep
South Field: 182 sheep

Test A

Chapter 6 Fractions

1 D

2 B

3 A

4 C

5 D

6 $\frac{2}{7}$

7 =

8 $\boxed{\frac{7}{10} \quad \left(\frac{7}{12}\right) \quad 1\frac{5}{12} \quad \left(1\frac{1}{2}\right)}$

9 $\frac{3}{13}, \frac{8}{13}, \frac{11}{13}, \frac{3}{3}$

10 $4\frac{1}{3}$

11 $8\frac{1}{4}$

12 $\frac{25}{12}$

13 $\boxed{4\frac{2}{3} \quad \left(\frac{30}{5}\right) \quad \left(\frac{59}{10}\right) \quad 5\frac{1}{6}}$

14 $3 \div 8 = \frac{3}{8}$

$\frac{3}{8}$ L

15 orange pumpkin

Test B

Chapter 6 Fractions

1 B

2 D

3 C

4 A

5 C

6 $\frac{5}{12}$

7 >

8 $7\frac{1}{4}$

9 $6\frac{1}{3}$

10 $\frac{22}{4}$ $\frac{8}{4}$ $\left(\frac{15}{6}\right)$ $\frac{14}{6}$ $\left(\frac{20}{8}\right)$

11 $1\frac{4}{16} = 1\frac{2}{8}$

$\frac{11}{5} = 2\frac{1}{5}$

12 $\frac{4}{12}$

13 $\frac{28}{7}$, $3\frac{6}{7}$, $\frac{20}{8}$, $2\frac{1}{3}$

14 $58 \div 5 = 11\frac{3}{5}$

$11\frac{3}{5}$ m

15 homework

Test A

Chapter 7 Adding and Subtracting Fractions

1 C

2 A

3 D

4 C

5 B

6 $\frac{25}{15} = \frac{5}{3} = 1\frac{2}{3}$

$1\frac{2}{3}$

7 $\frac{26}{18} + \frac{5}{18} = \frac{31}{18} = 1\frac{13}{18}$

$1\frac{13}{18}$

8 $\frac{10}{21} - \frac{6}{21} = \frac{4}{21}$

$\frac{4}{21}$

9 $\frac{7}{4} - \frac{5}{12} = \frac{21}{12} - \frac{5}{12} = \frac{16}{12} = \frac{4}{3} = 1\frac{1}{3}$

$1\frac{1}{3}$

10 $7\frac{2}{4} + 7\frac{3}{4} = 14\frac{5}{4} = 15\frac{1}{4}$

$15\frac{1}{4}$

11 $\frac{1}{2} + \frac{11}{18} = \frac{9}{18} + \frac{11}{18} = \frac{20}{18} = \frac{10}{9} = 1\frac{1}{9}$

$1\frac{1}{9}$ miles

12 $1\frac{2}{5} + 2\frac{2}{15} = 3\frac{6}{15} + \frac{2}{15} = 3\frac{8}{15}$

$3\frac{8}{15} + 3\frac{1}{5} = 6\frac{8}{15} + \frac{3}{15} = 6\frac{11}{15}$

$6\frac{11}{15}$ kg

13 $5\frac{3}{4} - 2\frac{1}{8} = 3\frac{6}{8} - \frac{1}{8} = 3\frac{5}{8}$

$3\frac{5}{8} - 1\frac{1}{2} = 2\frac{5}{8} - \frac{4}{8} = 2\frac{1}{8}$

$2\frac{1}{8}$ m

Test B

Chapter 7 Adding and Subtracting Fractions

1 C

2 B

3 D

4 B

5 A

6 $4\frac{1}{5} - \frac{2}{5} = 3\frac{6}{5} - \frac{2}{5} = 3\frac{4}{5}$

$3\frac{4}{5}$

7 $\frac{4}{8} + \frac{4}{8} + 1\frac{3}{8} = \frac{8}{8} + 1\frac{3}{8} = 1 + 1\frac{3}{8} = 2\frac{3}{8}$

$2\frac{3}{8}$

8 $6\frac{1}{16} - \frac{14}{16} = 5\frac{17}{16} - \frac{14}{16} = 5\frac{13}{16}$

$5\frac{13}{16}$

9 $\frac{10}{9} - \frac{3}{9} = \frac{7}{9}$

$\frac{7}{9}$

10 $1\frac{1}{5} + 2\frac{7}{10} = 3\frac{1}{5} + \frac{7}{10} = 3\frac{2}{10} + \frac{7}{10}$

$= 3\frac{9}{10}$

$3\frac{9}{10}$ lb

11 $\frac{5}{8} - \frac{1}{2} = \frac{5}{8} - \frac{4}{8} = \frac{1}{8}$

$\frac{1}{8}$ hr

12 $1 - \frac{2}{3} = \frac{3}{3} - \frac{2}{3} = \frac{1}{3}$

$\frac{1}{3} - \frac{2}{9} = \frac{3}{9} - \frac{2}{9} = \frac{1}{9}$

$\frac{1}{9}$

13 $3\frac{1}{6} - \frac{5}{12} = 2\frac{14}{12} - \frac{5}{12} = 2\frac{9}{12}$

$3\frac{1}{6} + 2\frac{9}{12} = 5\frac{2}{12} + \frac{9}{12} = 5\frac{11}{12}$

$5\frac{11}{12}$ m

Test A

Chapter 8 Multiplying a Fraction and a Whole Number

1 B

2 C

3 A

4 B

5 D

6 $\frac{2}{3} \times 5 = \frac{10}{3} = 3\frac{1}{3}$

$3\frac{1}{3}$

7 $\frac{5}{6} \times 9 = 5 \times \frac{\cancel{9}^3}{\cancel{6}_2} = 5 \times \frac{3}{2} = \frac{15}{2} = 7\frac{1}{2}$

$7\frac{1}{2}$

8 $12 \times \frac{3}{4} = \cancel{12}^3 \times \frac{3}{\cancel{4}_1} = 3 \times 3 = 9$

9

9 $\frac{4}{7} \times 21 = \frac{4}{\cancel{7}_1} \times \cancel{21}^3 = 4 \times 3 = 12$

12

10 $\frac{5}{6} \times 24 = \frac{5}{\cancel{6}_1} \times \cancel{24}^4 = 5 \times 4 = 20$

20 red apples

11 $30 - 16 = 14$

$\frac{14}{30} = \frac{7}{15}$

$\frac{7}{15}$

12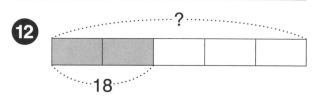

2 units ⟶ 18
1 unit ⟶ $\frac{18}{2} = 9$
5 units ⟶ $5 \times 9 = 45$

45 stickers

13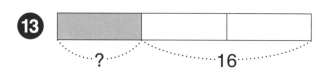

2 units ⟶ 16
1 unit ⟶ $\frac{16}{2} = 8$

8 peaches

Test B

Chapter 8 Multiplying a Fraction and a Whole Number

1 C

2 D

3 C

4 B

5 A

6 $8 \times \frac{5}{7} = \frac{40}{7} = 5\frac{5}{7}$

$5\frac{5}{7}$

7 $\overset{3}{\cancel{6}} \times \frac{1}{\underset{4}{\cancel{8}}} = \frac{3}{4}$

$\frac{3}{4}$

8 $\overset{}{\underset{2}{\cancel{\tfrac{3}{4}}}} \times \overset{25}{\cancel{50}} = \frac{3}{2} \times 25 = \frac{75}{2} = 37\frac{1}{2}$

$37\frac{1}{2}$

9 $20 + 15 + 5 = 40$

$\frac{5}{40} = \frac{1}{8}$

$\frac{1}{8}$

10 $\frac{7}{35} = \frac{1}{5}$

$\frac{1}{5}$

11 $\frac{7}{\underset{1}{\cancel{12}}} \times \overset{2}{\cancel{24}} = 14$

14 boys

12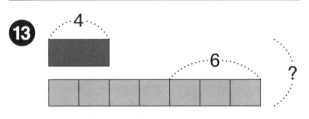

20 units ⟶ 160
1 unit ⟶ $\frac{160}{20} = 8$
9 units ⟶ $8 \times 9 = 72$
72 red and blue balls

13

3 units ⟶ 6
1 unit ⟶ $\frac{6}{3} = 2$
7 units ⟶ $7 \times 2 = 14$
$14 + 4 = 18$
$18

Test A

Chapter 9 Line Graphs and Line Plots

1. C

2. B

3. D

4. A

5. C

6. 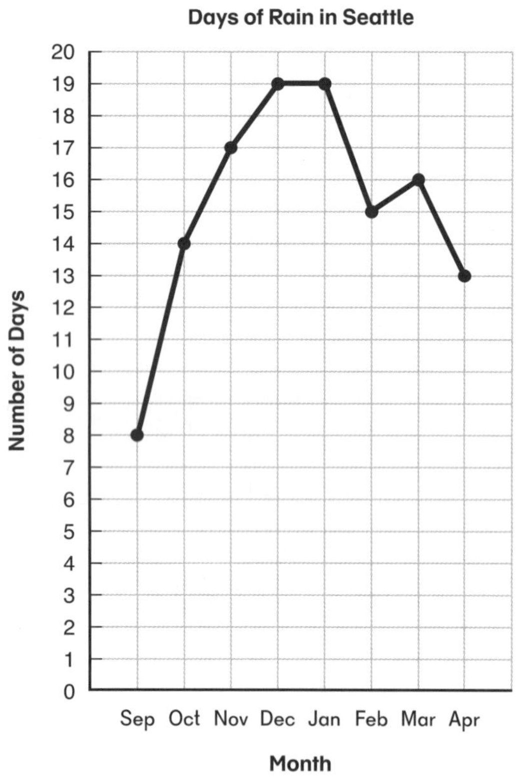

7. September

8. December and January

9. September and October

10. April

11. Runs or number of runs

12. 24 games

13. 7 runs

14. 8 games

15. 3 games had no runs and there was a total of 24 games.

$\frac{3}{24} = \frac{1}{8}$

$\frac{1}{8}$

Test B

Chapter 9 Line Graphs and Line Plots

1 C

2 A

3 B

4 D

5 B

6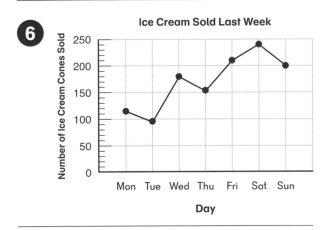

7 Number of ice cream cones sold

8 Saturday and Sunday

9 4 days

10 200 − 115 = 85
85 × 3 = 255
$255

11

12 26 bags

13 0 bags

14 $4 - 2\frac{1}{2} = 1\frac{1}{2}$
$1\frac{1}{2}$ lb

15 4 bags weighed $3\frac{1}{4}$ lb and there was a total of 26 bags.
$\frac{4}{26} = \frac{2}{13}$
$\frac{2}{13}$

Test A

Continual Assessment 2

1 D

2 B

3 B

4 A

5 B

6 A

7 A

8 B

9 C

10 A

11 52,799 (51,801) (52,499) 52,500 51,499

12
108 is a multiple of 6.	✓
78 is a multiple of 9.	
8 is a factor of 64.	✓

13 Factors of 24: 1, 2, 3, 4, 6, 8...
Factors of 72: 1, 2, 3, 4, 6, 8...

18 (24) 45 (72) 95

14 4,576

15 $\frac{2}{4} = \frac{1}{2}$

$\frac{2}{6} = \frac{1}{3}$

$\frac{1}{3}, \frac{1}{2}, \frac{5}{8}, \frac{8}{9}$

16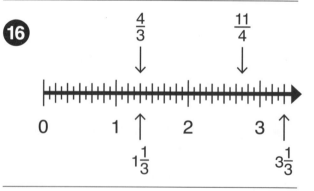

17 $8\frac{2}{6} - 3\frac{5}{6} = 5\frac{2}{6} - \frac{5}{6} = 4\frac{2}{6} - \frac{8}{6} = 4\frac{3}{6}$

$4\frac{1}{2}$ L

Test A

Continual Assessment 2

18 Books Read Over Summer

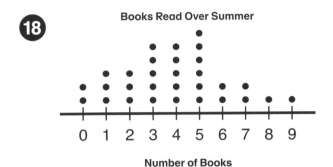

Number of Books

19 5 books

20 $\frac{5}{30} = \frac{1}{6}$

$\frac{1}{6}$

21 1,512 + 1,718 = 3,230
3,230 − 168 = 3,062
3,062 people

22 17 × 350 = 5,950
5,950 − 780 = 5,170
5,170 sheets of paper

23

Large	
Large	
Small	815
Small	
Small	
Small	
Small	

13,880

7 units ⟶ 13,880 − 815 − 815
= 12,250
1 unit ⟶ 12,250 ÷ 7 = 1,750
1,750 + 815 = 2,565
$2,565

24 After:

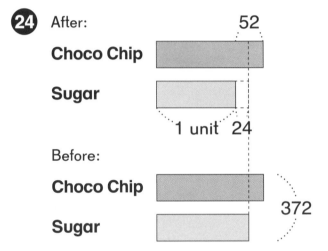

Total after: 372 − 24 = 348
2 units ⟶ 348 − 52 = 296
1 unit ⟶ 296 ÷ 2 = 148
148 + 52 = 200
148 + 24 = 172

200 chocolate chip cookies
172 sugar cookies

25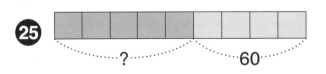

1 unit ⟶ 60 ÷ 4 = 15
5 units ⟶ 15 × 5 = 75
75 apple trees

Test B
Continual Assessment 2

1 D

2 C

3 B

4 C

5 A

6 D

7 A

8 D

9 A

10 C

11

| 52,799 | 93,621 | (95,000) | (123,500) | (149,889) |

12

All even numbers have at least one prime number as a factor.	✓
1, 4, and 9 have an odd number of factors.	✓
64 is a factor of 8.	

13 Factors of 24: 1, 2, 3, 4, 6, 8, 12…

Factors of 72: 1, 2, 3, 4, 6, 8, 9, 12…

Factors of 96: 1, 2, 3, 4, 6, 8, 12…

| 10 | 18 | (24) | 45 | (72) | 88 | (96) |

14 $\frac{9}{7}, 1\frac{1}{2}, \frac{13}{8}, \frac{13}{6}, 2\frac{3}{4}, \frac{16}{5}$

15 Use estimation.

$30 \times 80 = 2{,}400$

$50 \times 30 = 1{,}500$

$20 \times 60 = 1{,}200$

$80 \times 40 = 3{,}200$

| 32 × 81 | 51 × 27 | 18 ✗ 63 | (78 × 36) |

Test B

Continual Assessment 2

16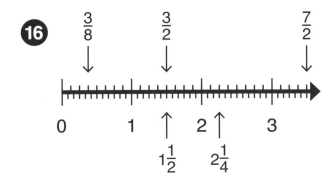

17 $9\frac{3}{4} - 3\frac{1}{2} = 6\frac{3}{4} - \frac{2}{4} = 6\frac{1}{4}$

$6\frac{1}{4} - 2\frac{1}{4} = 4$

4 kg

18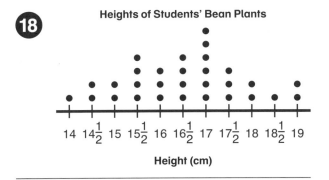

Heights of Students' Bean Plants
Height (cm)

19 9 bean plants

20 $\frac{8}{30} = \frac{4}{15}$

$\frac{4}{15}$

21 Trail B: 425 + 2,245 = 2,670
Trail C: 2,670 − 187 = 2,483
2,483 − 2,245 = 238
238 m

22 24 × 750 = 18,000
3 × 750 = 2,250
18,000 − 2,250 − 300 = 15,450
15,450 mL

23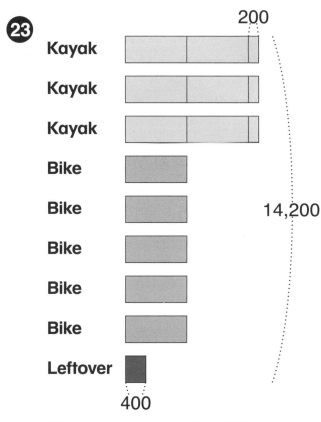

11 units ⟶ 14,200 − 200 − 200 − 200 − 400 = 13,200
1 unit ⟶ 13,200 ÷ 11 = 1,200
1,200 + 1,200 + 200 = 2,600
$2,600

Test B

Continual Assessment 2

24 Before:

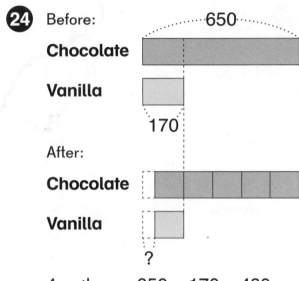

After:

?

4 units ⟶ 650 − 170 = 480
1 unit ⟶ 480 ÷ 4 = 120
170 − 120 = 50
50 × 2 = 100
100 pints

25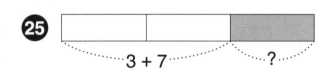

2 units ⟶ 10
1 unit ⟶ 10 ÷ 2 = 5
3 units ⟶ 3 × 5 = 15
15 miles